Mohammed Ait El Fqih

# Corrected non-destructive testing (NDT) exercises

Mohammed Ait El Fqih

# Corrected non-destructive testing (NDT) exercises

NDT exercises

ScienciaScripts

**Imprint**



Cover image: www.ingimage.com

This book is a translation from the original published under ISBN 978-3-8416-1320-2.

Publisher:
Sciencia Scripts
is a trademark of
Dodo Books Indian Ocean Ltd. and OmniScriptum S.R.L publishing group

120 High Road, East Finchley, London, N2 9ED, United Kingdom
Str. Armeneasca 28/1, office 1, Chisinau MD-2012, Republic of Moldova, Europe
Managing Directors: Ieva Konstantinova, Victoria Ursu
info@omniscriptum.com

Printed at: see last page
**ISBN: 978-620-8-50017-7**

# *Table of contents*

# Foreword

**Dear students,**

We are pleased to present this collection of corrected exercises dedicated to Non-Destructive Testing (NDT). Designed for the Années Préparatoires Intégrées cycle, this book offers you a structured, in-depth introduction to the essential methods of NDT, a fundamental field in mechanics, aeronautics and many other industrial sectors where safety and reliability are paramount.

In this collection, you will explore the main NDT procedures, including ultrasonic testing, eddy current, radiography and thermography. These methods will enable you to learn how to identify, analyze and interpret defects without damaging materials or structures. The varied exercises and detailed answer keys will guide you step by step in acquiring these skills, consolidating your theoretical knowledge while giving you the tools you need to apply them in practical contexts.

Non-destructive testing is a discipline at the heart of many industrial advances, guaranteeing product safety and quality while optimizing costs and resources. A good understanding of these concepts will open up a wide range of opportunities in innovative sectors. This book is designed to support you in this learning process, with a practical approach supported by real-life examples and solved exercises to bring the concepts to life and make them accessible.

We hope this guide will become a valuable ally in your studies, preparing you for the technical and intellectual challenges you will encounter in the dynamic field of NDT.

We wish you a rewarding career in the world of Non-Destructive Testing, and every success in your future academic and professional projects.

***Prof. Mohammed AIT EL FQIH***
***Branch: Industrial Mechanical Systems Engineering***

# Part 1:
# Visual Control

## Exercise 1: Origin of discontinuities

Or discontinuities introduced into a raw material (foundry), as shown in figure 1.

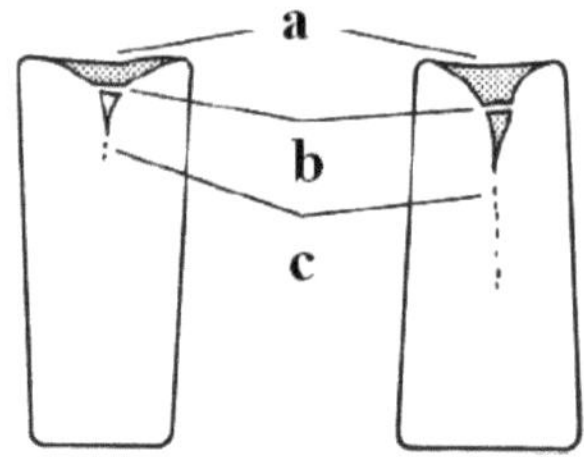


***Figure 1***

1- Name zones a, b and c.

2- Specify the nature of each type of discontinuity shown in Figure 1.

3- Roughly suggest the appropriate control method(s) for each type of discontinuity.

**Correction:**

1- Name zones a, b and c.

a: Retassure (temperature gradient => a crack in the form of a retassure)

b: Bypass

c: Axial shrinkage

2- Specify the nature of each type of discontinuity shown in Figure 1.

Shrinkage: surface discontinuity

Bridging: non-through discontinuity

Axial shrinkage: volume discontinuity

3- Roughly suggest the appropriate control method(s) for each type of discontinuity.

Shrinkage: surface discontinuity => Visual, Penetrant Testing

Bypass: non-through discontinuity => Magnetoscopy

Axial shrinkage: volume discontinuity => Eddy current or ultrasound.

## Exercise 2: Checking a welded part

1- Specify the types of discontinuity likely to be present in the molding process.

2- Or discontinuities introduced during welding of a mechanical part (figure 2).

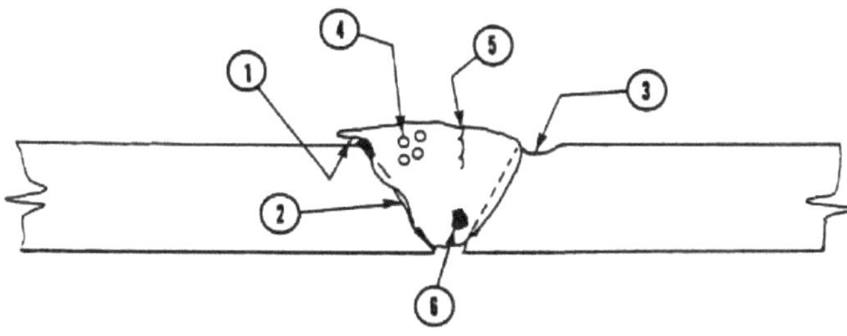

*Figure 2*

2-a Name zones 1 to 6.

2-b Specify the nature of each type of discontinuity shown in Figure 2.

3- Roughly suggest the appropriate control method(s) for each type of discontinuity.

4- What is the damage introduced in service during welding?

**Correction:**

2-a Name zones 1 to 6.

1: overlap

2: Blow

3: Gutter

4: Porosity

5: Retassure

6: Slag inclusion

2-b Specify the nature of each type of discontinuity shown in Figure 2.

1: Overlap: Subcutaneous surface

2: Blow: Volumetric

3: Gutter: Surface area

4: Porosity: Volumetric or surface, non-throughput

5: Shrinkage: Open area

6: Slag inclusion: Volumetric

3- Roughly suggest the appropriate control method(s) for each type of discontinuity.

1: Overlap: Subcutaneous surface => Visual

2: Blowing: Volumetric => Eddy current or ultrasonic

3: Gutter: Surface => Visual

4: Porosity: Volumetric or non-through surface => Magnetoscopy

5: Shrinkage: Open surface => Penetrant testing

6: Slag inclusion: Volumetric => Eddy current or ultrasonic

Welding, despite its effectiveness in joining various materials, can introduce several types of damage and defects into a structure during its use in service. Here are the main types of damage associated with welding:

Cracking :

Hot cracks: These appear during weld solidification, due to high thermal stresses, often accentuated by high levels of impurities (sulfur, phosphorus).

Cold cracks (delayed cracks): These cracks form after welding during cooling, often due to high residual stresses, hydrogen present in the base or filler metal, or insufficient ductility.

Residual stress :

During welding, the base metal and filler metal undergo thermal cycles which induce residual stresses. These stresses, if high, can weaken the strength of the welded joint, making it vulnerable to stress cracking or fatigue in service.

Deformation :

Heating and cooling cycles often cause deformations in the welded structure, which can induce stresses and compromise the alignment or dimensions of the assembly. This phenomenon is common in large parts.

Microcracks and porosity :

Corrosion :

The difference in chemical composition and microstructure between the weld metal and the base metal can make the welded area more susceptible to corrosion.

Heat-affected zones, in particular, can be weak points, especially in aggressive environments (humidity, salts, chemicals).

Awareness and fragility :

Particularly in stainless steels, welding can cause precipitation of chromium carbides at grain boundaries, making the thermally-affected zone vulnerable to intergranular corrosion (sensitization).

Hydrogen embrittlement is another phenomenon that can occur, especially in high-strength steels, where the absorption of hydrogen during welding embrittles the material in service.

Alteration of the microstructure :

Welding thermal cycles modify the microstructure of the base metal and weld metal. In some materials, this can lead to changes in mechanical properties (hardness, ductility) and affect in-service strength, making the material more susceptible to cracking.

Such damage varies according to material type, welding process and service conditions. Non-destructive testing practices, such as ultrasonic or X-ray testing, can detect these defects before they propagate in service.

## Exercise 3: Throat size control

During the stirring process, you have to check the quality of the weld without compromising the quality of 3 parts (see figure 3).

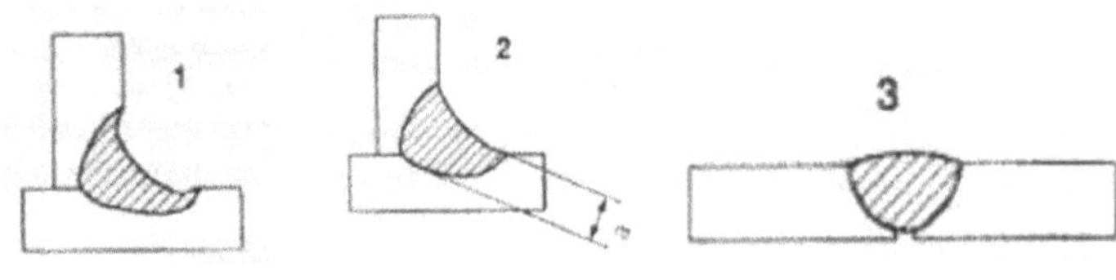


***Figure 3***

What is the nature of the discontinuity present on each type of weld in figure 3?

**Correction:**

We start by finding the throat size for each weld.

We use the relationship$a_{Th} = e.0{,}7$ for welding in T

If$a_{mesurée} > a_{Th}$ we validate the part

If$a_{mesurée} < a_{Th}$ the part is not validated

Conclusion:

Part 1 invalid

Valid part 2

For flat welding, we use the relationship :

If$a_{mesurée} > a_{Th}$ we validate the part

If$a_{mesurée} < a_{Th}$ the part is not validated

With $a_{Th} = e$

Conclusion:

Invalid part 3

## Exercise 4: String width control

Consider two metal sheets of equal thickness s, welded over a length L, and subjected to tension with a force F.

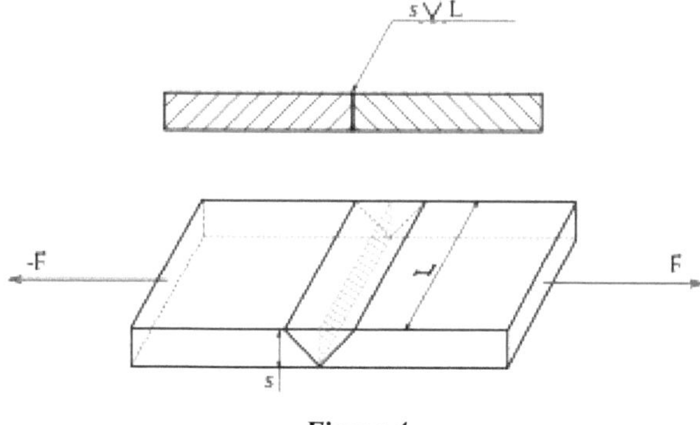


***Figure 4***

Find :

1- The area of the throat plane, hatched in grey on the figure.

2- The normal stress σ subjected to the throat plane.

3- The weld strength condition.

4- Minimum cord length

Application:

S235 steel plate (Re = 235 MPa) with thickness s = 5 mm, subjected to force F = 5,000 N and safety factor k = 2

**Correction:**

1- $A = L.e$

2- $\sigma = F/_{A} = F/_{L.e}$

3- $\sigma \leq R_e/_{k}$

Where k is the safety coefficient.

4- - $\sigma \leq R_e/_{k} \Rightarrow F/_{L.e} \leq R_e/_{k} \Rightarrow L \geq F.k/_{R_e.e}$

# Part 2:
# Liquid penetrant inspection

## Exercise 5: Penetrant testing steps

1- Name the different steps (from a to f) in the penetrant inspection of a mechanical part (figure 5).

2- Detail each step.

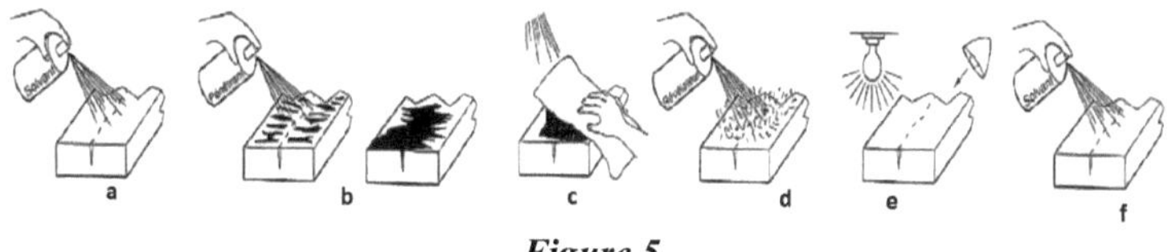


***Figure 5***

3- What are the types of developer s.

4- What are the advantages and limitations of the dye penetrant inspection technique?

5- What are the fields of application for penetrant testing?

**Correction:**

1- and 2- Penetrant testing is a non-destructive testing method used to detect open surface defects, such as cracks or porosities, on non-porous materials. Here are the main steps in the process, from a to f :

a. Surface preparation

Thoroughly clean the surface of the part to remove any dirt, oil, grease or oxidation that might prevent the penetrant from entering the defects. Solvents, degreasers or mechanical cleaning can be used to obtain a clean surface.

b. Penetrant application

Apply a penetrating liquid (usually brightly colored or fluorescent) to the surface of the part. This liquid is designed to infiltrate open surface discontinuities by capillary action. It is left to act for a given period of time (penetration time), so that the liquid penetrates well into the defects.

c. Removal of excess penetrant

After the penetration time, remove excess penetrant from the surface. This step must be carried out carefully to avoid removing penetrant that has infiltrated the defects.

The cleaning method depends on the type of penetrant used (water-based, water-soluble or post-emulsifiable).

d. Application of developer

Apply a developer to the surface to bring out the penetrant remaining in the defects. The developer absorbs the penetrant and brings out the indications in the form of visible marks, enabling discontinuities to be located and visualized.

e. Observation and interpretation

Examine the part under appropriate light (white for colored penetrant, ultraviolet for fluorescent penetrant) to spot indications. Defects appear as colored or fluorescent marks, and their size and shape are analyzed to assess the extent of discontinuities.

f. Final cleaning

Clean the part to remove developer and penetrant residues, making sure to leave the surface in its original condition. This is especially important if the part is to be put back into service.

These steps enable open surface defects to be effectively identified, and are essential for guaranteeing the quality and safety of inspected mechanical parts.

3- Developers used in penetrant testing are classified according to their form and application. Here are the main types of developer:

Dry developers

These are fine powders, often applied as a dry powder directly to the surface of the part. These developers are mainly used with fluorescent penetrants and are well suited to dry, non-porous parts. They are applied by dusting or spraying, and are particularly suitable for parts with complex shapes.

Wet developers in aqueous suspension

These developers are powders suspended in water with wetting agents to improve adhesion to the part surface. They are applied by spray or immersion. Once the water has evaporated, the powder forms a thin film, absorbing the penetrant and revealing discontinuities. These developers often require controlled drying to avoid excess humidity.

Wet developers in non-aqueous suspension

This type of developer consists of fine particles suspended in a volatile solvent, often applied by pressure spraying. The solvent evaporates rapidly, leaving a fine powdery film. These developers are particularly effective for colored penetrants, and ensure high sensitivity in defect detection.

Water-soluble developers

Water-soluble developers are applied as an aqueous solution, forming a white film after drying. They are often used with fluorescent penetrants, and give good contrast of indications under ultraviolet light. They require rigorous drying to prevent traces of water from affecting the quality of the indication.

Peel-off film developers

These less common developers are applied in liquid form to create a continuous film on the surface of the part. Once dry, the film can be removed, taking with it the penetrant trapped in the defects, facilitating inspection. This type is mainly used for hard-to-reach surfaces or specific applications.

Each type of developer is chosen according to the type of penetrant, the nature of the part and the test conditions. Dry developers and non-aqueous suspension developers are generally preferred for their ease of use and adaptability.

4- Dye penetrant inspection is a popular method in the field of non-destructive testing (NDT) for detecting open surface defects. Here are the main advantages and limitations of this technique:

Advantages of dye penetrant inspection

Simplicity and low cost

The method is relatively simple to implement and requires no expensive equipment. This makes it a cost-effective solution for surface inspections.

High sensitivity for surface defects

It enables the detection of very fine discontinuities (cracks, porosities) on the surface, which other methods might not detect as easily.

Versatility

Penetrant testing can be applied to a wide variety of non-porous materials, including metals, plastics, ceramics and composites.

Adaptability

This method can be used on parts of all shapes and sizes, even with complex geometries. It is suitable for both on-site and workshop inspections.

Easy-to-interpret visual results

The indications (colored or fluorescent marks) produced by the developer are visible to the naked eye, making it easier to detect and interpret defects.

Quick to apply

Penetrant testing is quick to perform and offers immediate results, enabling rapid intervention on the tested parts if necessary.

Limits of dye penetrant testing

Limited to open surface defects

This technique cannot detect internal or subsurface defects, such as buried cracks or weld defects that do not propagate to the surface.

Incompatibility with porous materials

Porous materials, such as certain ceramics or concrete, absorb the penetrant in an uncontrolled way, making flaw detection ineffective.

Requires rigorous cleaning

Careful surface preparation is essential for reliable results. Any contamination (oil, grease, dust) can prevent the penetrant from penetrating defects properly.

Sensitivity to environmental conditions

Temperature and humidity conditions can affect penetrant and developer performance. Extreme conditions can alter the quality of results.

Risk of false indications

Poorly cleaned penetrant or developer residues can create false indications, making interpretation difficult or imprecise.

Handling chemicals

The products used (penetrants, solvents and developers) can be chemical and require health, safety and environmental precautions. They can also pose waste management problems.

5- Penetrant testing is widely used in various industrial sectors due to its ability to detect surface defects efficiently. The main fields of application are

Aeronautics and aerospace

In these sectors, safety and reliability are paramount. Penetrant testing is used to inspect critical parts such as turbine blades, disks, rotor blades and fuselage components for cracks, porosity or other discontinuities that could affect aircraft performance and safety.

Automotive industry

Dye penetrant testing is commonly used to inspect mechanical components such as engine blocks, cylinder heads, drive shafts and chassis parts. It ensures the absence of surface defects that could compromise component durability.

Energy industry

In the energy sector, and more particularly in nuclear, thermal and wind power plants, penetrant testing is used to check parts in critical service, such as piping, welds and pressure vessels. It enables components to be checked under extreme conditions, minimizing the risk of leakage or rupture.

Metallurgy and foundry

In foundry and forge shops, dye penetrant testing is used to check the quality of cast, forged or machined parts (e.g. connecting parts, complex castings, etc.). This enables casting faults, such as cooling cracks and surface blowholes, to be detected before the parts are put into service.

Shipbuilding

This process is essential for inspecting ship hulls, propellers, welds and other parts exposed to corrosive environments and high pressures. It ensures the reliability of welds and components subject to high stress.

Oil and gas industry

Penetrant testing is used to inspect storage tanks, pipelines, welds and other critical equipment exposed to corrosive, high-pressure fluids. This inspection guarantees the safety of installations and prevents the risk of industrial accidents.

Railway industry

Dye penetrant testing is used to inspect components such as wheels, axles and suspension elements. It is used to detect fatigue cracks and other surface defects that could affect the safety of rail vehicles.

Industrial maintenance

Penetrant testing is commonly used in the preventive and corrective maintenance of industrial machinery to check parts subject to high stress, such as shafts, gears and bearing housings, and to ensure their integrity.

Nuclear industry

In this safety-critical sector, penetrant testing is used to inspect welds, containment tanks and piping. It enables the rapid detection of cracks and surface defects that could compromise the safety of installations.

## Exercise 6: Possibilities and limitations of dye penetrant testing.

- Interpret the dye penetrant inspection possibilities and limitations diagram (figure 6).

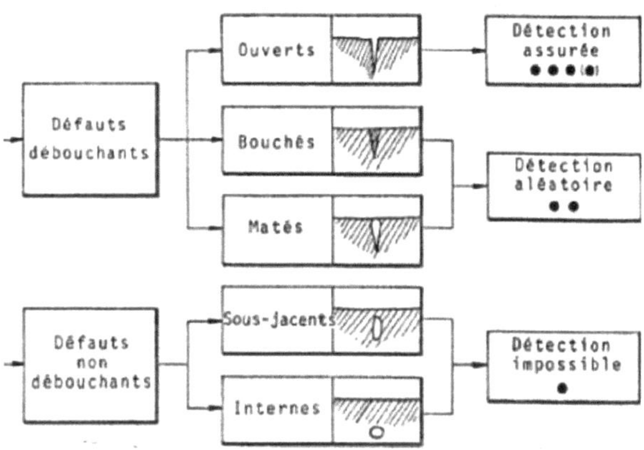


*Figure 6*

**Correction:**

This diagram illustrates the possibilities and limitations of penetrant testing according to the type of defect present in a part. Here is an interpretation of the different categories:

1. Open defects

Open defects: These defects are open at the surface, allowing the penetrant to penetrate easily. The diagram indicates that detection is assured for this type of defect, as they are well exposed at the surface. Examples: cracks or open porosity.

Plugged defects: These defects are located on the surface, but their opening is partially obstructed by contaminants or hard materials. The diagram indicates random detection, as success depends on the penetrant's ability to penetrate through the obstruction.

Matt defects: These defects are present on the surface, but their opening is very narrow due to the partial closure of the edges. Here too, detection is uncertain, as the penetrant may have difficulty penetrating these closed areas.

2. Non-through defects

Underlying defects : These defects are close to the surface but not directly exposed. Penetrant testing cannot reach these defects, and detection is indicated as impossible. Examples: sub-surface microcracks.

Internal defects: These defects are located deep inside the part and do not communicate with the surface at all. Detection is impossible for these defects because of their inaccessibility to the penetrant.

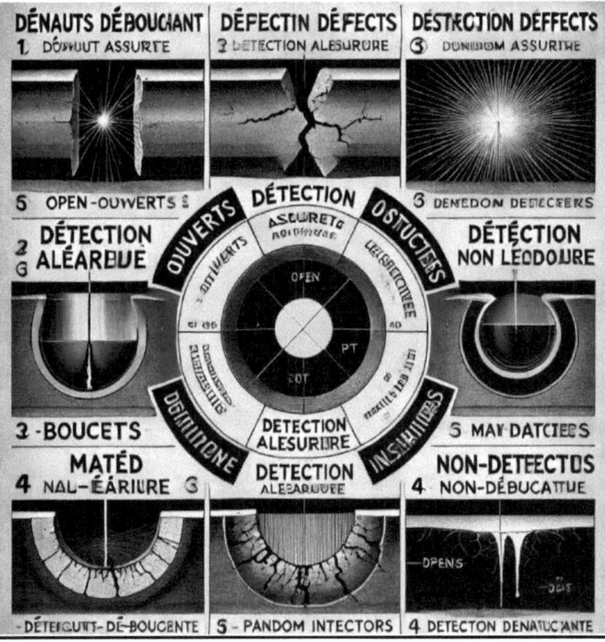

## Exercise 7: Classification of dye penetrant inspection processes.

To characterize a given penetrant testing process, the following ISO classification is used (Figure 7).

1- Identify, according to ISO 3452-1, the product family comprising :

a- a fluorescent penetrant, sensitivity N°4, which is removed with a hydrophilic emulsifier, then revealed with dry powder;

b- a coloured penetrant, sensitivity N°3, which is removed with a liquid solvent, then revealed by special application.

2- Conversely, find the penetrant testing process identified in ISO 3452-1 as : IIICe2, IIAd3 and IEa4.

| | ISO 3452-1 |
|---|---|
| | Type de pénétrant |
| Fluorescent | I |
| Coloré | II |
| Mixte (fluorescent et coloré) | III |
| | Méthode d'élimination du pénétrant |
| Eau | A |
| Emulsifiant lipophile | B |
| Solvant (liquide) | C |
| Emulsifiant hydrophile | D |
| Eau et solvant | E |
| | Forme du révélateur |
| Sec | a |
| Hydrosoluble | b |
| Suspension dans l'eau | c |
| A base de solvant (non aqueux pour type I) | d |
| A base de solvant (non aqueux pour types II et III) | e |
| Application spéciale | f |

***Figure 7***

**Correction:**

1- Identify, according to ISO 3452-1, the product family comprising :

a- a fluorescent penetrant, sensitivity N°4, which is removed with a hydrophilic emulsifier, then revealed with dry powder;

$$I\,B\,a\,4$$

b- a colored penetrant, sensitivity N°3, which is removed with a liquid solvent, then revealed by special application.

***II C f 3***

2- Conversely, find the penetrant testing process identified in ISO 3452-1 as :

- ***III C e 2***

A mixed penetrant, sensitivity N°2, which is removed with a liquid solvent, then revealed by solvent.

***II A d 3***

A colored penetrant, sensitivity N°3, which is removed with water, then revealed by solvent.

## Exercise 8: Choice of penetrant

1- Products used as penetrants must have a number of essential characteristics. Name these characteristics.

2- Consider a drop of two liquids whose wettability is shown in figure 8.

***Figure 8***

Discuss the wettability of the two drops as a function of angleθ. Deduce the type of surface

3- Discuss the effect of dynamic viscosity on the residence time of the different liquids in Table 1. Which is the best penetrant.

Reminder: $t = A.\mu$ with$A = 2.l^2/r.\cos(\theta).\gamma$; For the suite$A = 1$ .

$V = \frac{\mu}{\rho}$; $V$: *Viscosité cinématique*; $\mu$: *Viscosité dynamique*; $\rho$: *masse volumique*

| Type of liquid | Surface tension (mN/m) | Density ($kg/dm^3$) | Dynamic viscosity (Pa.s) | Kinematic viscosity ($mm^2/s$) |
|---|---|---|---|---|
| Water | 73 | 1 | 0,001 | |
| Alcohol | 23 | 0,78 | 0,012 | |
| Oil | 31 | 0,89 | 0,09 | |
| Ardrox 985 P3 penetrant | 27,5 | 0,95 | | 9,3 |
| Water-based penetrant Fluxo P502 | 25 | 0,99 | | 10,5 |

***Table 1***

**Correction:**

1- The products used as penetrants in penetrant testing must have several essential characteristics to guarantee reliable and effective defect detection. Here are the main ones:

Good wettability

The penetrant must be able to spread evenly over the surface of the part to penetrate defects easily. This requires low surface tension to thoroughly wet even the smallest cracks.

High penetration capacity

The product must be able to penetrate the smallest open surface discontinuities to reach all types of defects, even narrow cracks.

Clear visibility

The penetrant must offer optimum visibility, especially after the developer has been applied. Colored or fluorescent penetrants are commonly used to ensure good distinction between the defect and the surface of the part.

Easy to clean

After the penetration period, the product must be easy to remove from the surface without leaving residues that could interfere with the interpretation of results.

Compatibility with materials

The penetrant must not react chemically with the material of the part inspected, nor cause corrosion or surface deterioration.

Chemical stability

The product must be stable and retain its properties over time, even when exposed to variations in temperature, light or humidity. This prevents loss of effectiveness during storage.

Low toxicity and environmental safety

To minimize risks to operators and the environment, the penetrant must be low-toxicity, non-flammable and ideally biodegradable or recyclable.

Adequate drying time

The penetrant must be able to dry quickly enough after application, but without compromising the quality of penetration into defects, to facilitate the cleaning and inspection stages.

2- Wettability and Contact Angle $\theta$ :

***Figure 9***

The wettability of a liquid on a solid surface depends on the contact angle θ, which is the angle formed between the drop surface and the solid surface at the contact line.

If the angle θ is small (typically ):$\theta < 90 \circ$

The drop spreads more widely over the surface, indicating good wettability. In this case, the liquid is said to be hydrophilic or oleophilic, depending on the nature of the liquid (hydrophilic for water, oleophilic for oils). A surface with a low contact angle is often referred to as hydrophilic.

If the angle θ is large (typically ):$\theta > 90 \circ$

The drop forms a more rounded ball, indicating poor wettability. In this case, the liquid is hydrophobic or oleophobic. A surface with a large contact angle is called hydrophobic.

Discussion of the wettability of the Two Drops

Based on the contact angles observed for the two drops, we can draw conclusions about the wettability of the surface for each liquid:

If one drop has a low contact angle and the other a high angle :

The surface is probably hydrophilic for the low-angle liquid (which spreads better) and hydrophobic for the high-angle one. This may indicate that the surface is polar or non-polar, influencing the way different liquids adhere to it.

If the two drops have a small angle :

The surface is generally hydrophilic or oleophilic, compatible with different types of liquids (water, oil). This may indicate that the surface is polar, or has been treated to enhance liquid adhesion.

If the two drops have a high angle :

The surface is hydrophobic or oleophobic, suggesting a low-polarity surface or one with a non-stick coating, such as treatments to make it resistant to water or oils.

4- the effect of dynamic viscosity on the residence time of the different liquids in Table 1. Which is the best penetrant.

Reminder: $t = A.\mu$ with$A = 2.l^2/r.\cos(\theta).\gamma$; For the suite$A = 1$ .

$V = \frac{\mu}{\rho}; V: Viscosité\ cinématique;\ \mu: Viscosité\ dynamique;\ \rho: masse\ volumique$

| Type of liquid | Surface tension (mN/m) | Density (kg/dm$^3$) | Dynamic viscosity (Pa.s) | Kinematic viscosity (mm$^2$/s) |
|---|---|---|---|---|
| Water | 73 | 1 | 0,001 | |
| Alcohol | 23 | 0,78 | 0,012 | |
| Oil | 31 | 0,89 | 0,09 | |
| Ardrox 985 P3 penetrant | 27,5 | 0,95 | | 9,3 |
| Water-based penetrant Fluxo P502 | 25 | 0,99 | | 10,5 |

***Table 2***

# Part 3:
# Magnetic particle inspection

## Exercise 9: Magnetic particle inspection

Or a mechanical part controlled by magnetic particle inspection (Fig. 10).

1- What do the numbers 1 to 5 correspond to?

2- What type of fault can this control mode detect?

3- You are asked to inspect a mechanical part using magnetic particle inspection (Fig. 11).

3-1 Define the various steps to be taken to carry out this inspection.

3-2 Interpret each sub-figure 11.

4- Consider two parts (A: hard steel and B: mild steel) to be inspected by magnetic particle inspection. Their hysteresis cycles are given in figure 12.

4-1 Interpret figure 3.

4-2 Which part will be efficiently controlled? Justify your answer.

4-3 If the temperature of these two steels is 790°C. Will the test still be valid? Why?

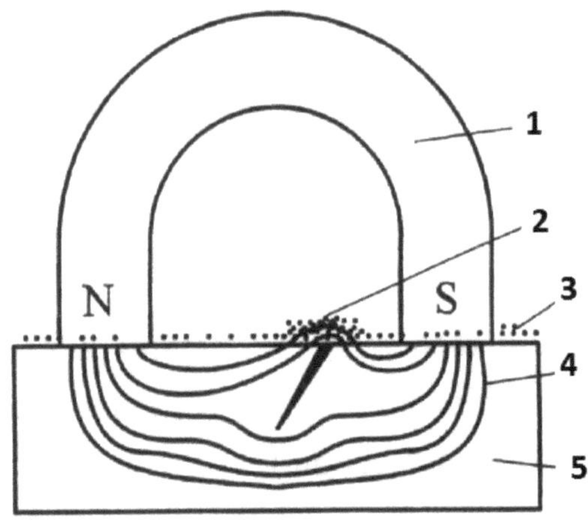


*Figure 10*

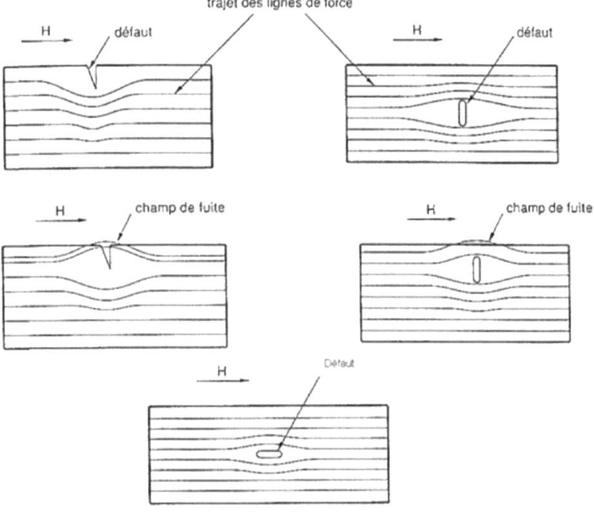


*Figure 11*

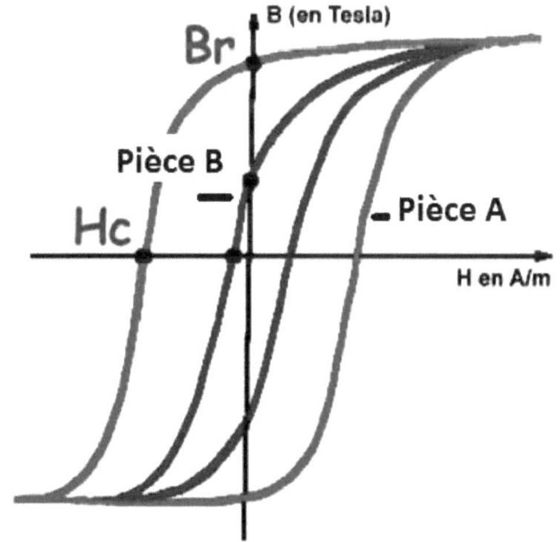


*Figure 12*

**Correction:**

1-

1: Permanent magnet,

2: Field line leak,

3: Tracer,

4: Field lines,

5: part to be analyzed

2- Magnetic particle inspection (MPI) is a non-destructive testing (NDT) method used primarily to detect surface and underlying near-surface defects in ferromagnetic materials. It is particularly effective in identifying the following types of defect:

a) **Surface cracks**: such as those caused by fatigue, stress or wear. These cracks are often visible on the surface, but can be very fine and difficult to detect with the naked eye.
b) **Pitting and porosity**: small cavities or holes in the surface, often caused by corrosion or manufacturing defects.
c) **Laminations**: poorly welded or poorly formed layers that delaminate, often appearing near the surface.
d) **Defective welds**: MT can identify surface weld defects, such as cracks or slag inclusions.
e) **Foreign inclusions**: the presence of non-metallic inclusions or foreign bodies in the material, which can affect the integrity of the structure.

3-

3-1 :Magnetic particle testing requires meticulous preparation and precise monitoring to guarantee reliable results. Here are the key steps for carrying out a magnetic particle inspection:

**1. Surface preparation**

a) **Cleaning**: The surface of the part must be clean, free of oil, grease, dust, paint or other contaminants that could interfere with fault detection.
b) **Mechanical preparation** (if necessary): If the surface is rough or has stubborn residues, light sanding or stripping can be carried out.

**2. Part magnetization**

- **Choice of magnetization method**: Depending on the shape of the part, its size and the type of defect to be investigated, different magnetization techniques can be used, such as :
    - **Circular current**: for cylindrical parts, to detect transverse defects.
    - **Longitudinal current**: for long parts, to detect cracks parallel to the part axis.
- **Magnetic field application**: A magnetic field is applied to the part to polarize the magnetic particles around potential defects.

**3. Application of Magnetic Particles**

- **Dry powder application**: Magnetic particles in the form of dry powder are sprinkled onto the surface of the part. They accumulate around defects, creating a visible indication.
- **Wet particle application**: In some cases, particles are applied as an aerosol or suspension in a liquid. This method is often used for complex-shaped parts.

**4. Fault observation and detection**

- **Lighting**: Inspection must be carried out under adequate lighting. If fluorescent particles are used, a UV lamp is required to highlight defects.
- **Visual observation**: Magnetic particles accumulate around defects, forming visible patterns that indicate their location and shape.
- **Interpreting indications**: Carefully examine visible indications to assess the size, orientation and nature of the defect.

**5. Part demagnetization (if necessary)**

- **Demagnetizing**: After inspection, it is often necessary to demagnetize the part to remove any residual fields that could affect its future use or mechanical properties.

**6. Final cleaning**

- **Removal of magnetic particles**: A final cleaning is carried out to remove all magnetic particles and suspending agents.
- **Cleanliness check**: Ensure that the part is in its original condition, ready for use or further inspection if required.

**7. Results documentation**

- **Inspection report**: Document the results of the inspection, including indications observed, their location, size, and any recommendations for further repair or evaluation.

**Checklist de Contrôle Non Destructif - Magnétoscopie**

1. Informations Générales

- Date du contrôle : ______
- Lieu d'inspection : ______
- Nom de l'inspecteur : ______
- Qualification de l'inspecteur : ______
- Référence de la pièce : ______
- Matériau de la pièce : ______
- Description de la pièce : ______
- Méthode de magnétoscopie utilisée : Poudre sèche / Particules humides

2. Préparation de la Surface

- Surface propre (huile, graisse, peinture enlevée)
- Préparation mécanique (ponçage ou décapage si nécessaire)

3. Paramètres de Magnétisation

- Type de courant : Courant alternatif / Courant continu
- Technique de magnétisation : Circulaire / Longitudinale
- Intensité de courant : ______ A
- Position des électrodes : ______

4. Application des Particules Magnétiques

- Type de particules : Sèches / Humides
- Couleur/Fluorescence : ______
- Mode d'application : Poudre manuelle / Spray / Autre

5. Observation des Indications

- Éclairage adéquat (lumière blanche / UV)
- Indication n°1 : Localisation ______ Taille ______ Type ______
- Indication n°2 : Localisation ______ Taille ______ Type ______

6. Démagnétisation de la Pièce (si applicable)

- Procédure de démagnétisation effectuée : Oui / Non
- Résidu magnétique mesuré : ______ Gauss

7. Conclusion et Recommandations

- Résumé des résultats : ______
- Recommandations : Réparation / Remplacement / Surveillance accrue
- Statut de la pièce : Conforme / Non conforme

8. Documentation et Archivage

- Photos des indications en annexe : Oui / Non
- Signatures : Inspecteur ______ Superviseur ______

**Archiving**: Keep photos and notes in the inspection report for future reference, to ensure traceability and track defects.

When the temperature of these two steels reaches 790°C, control becomes inapplicable, as this temperature exceeds the Curie temperature of 750°C. At this critical temperature, the steel loses its magnetic properties, which significantly affects its behavior in applications requiring specific magnetic characteristics. Consequently, any evaluation or test based on magnetic properties at this temperature will not be able to provide reliable results. It is therefore essential to closely monitor temperatures when using these steels to ensure their performance and integrity in the applications envisaged.

# Part 4: Eddy current control

## Exercise 10: Losses in laminated metal

A metal is laminated into identical sheets of rectangular parallelepiped shape (Fig. 13), of dimensions ,$ab$ and$h$ ; the reference frame$(Oxyz)$ , linked to a sheet, has as its origin the center$O$ of the parallelepiped and as its axes the axes of symmetry (figure 1.a - see appendix) ; the conductivity of the sheet is noted .$\sigma$

The metal is subjected to a low-frequency sinusoidal magnetic field whose vector

$$\vec{B} = B_m . sin(\omega t) . \vec{e}_z$$

Where$B_m$ is a positive scalar independent of time and space coordinates.

Under these conditions, the metal is the seat of eddy currents: we propose evaluate the power dissipated by these currents in a metal sheet. Taking into account the direction of$\vec{B}$ and the shape of the sheet, we consider the following model, where currents flow in sheet metal elements such as the one shown in figure 1.b :

- Its dimensions are:$b$ in the direction ,$Ox2y$ in the direction$Oy$ and$h$ in the direction .$Oz$
- The conductor cross-section is .$h. dy$

After calculation, the power dissipated by the Joule effect is given by :

$$p_J = \gamma \frac{a^2}{24} B_m^2 \omega^2 = \gamma \frac{a^2}{6} B_m^2 \pi^2 f^2$$

1- Indicate how the power calculated in the previous question can be minimized for a given magnetic field.

2- To obtain the minimum power expenditure, choose between aluminum and steel, and evaluate the corresponding dissipated power density.

3- Is it possible to detect cracks inside each metal?

4- To what depth can cracks be detected on each metal?

Data :

$\sigma$ (Al)= $3.7.10^7$ U.S.I ;$\sigma$ (steel)= $3.17.10^5$ U.S.I

$B_m$=0.1 T;$\omega = 100\pi$ rad.$s^{-1}$; a= 2mm.

$\mu_r(Acier) = 200$ ; $\mu_r(Al) = 1,0007$ ; $\mu_0 = 4\pi . 10^{-7}\ H.m^{-1}$

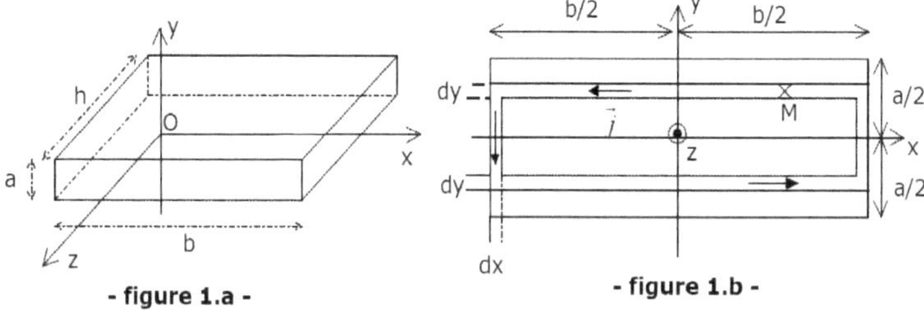


***Figure 13 (1.a -1. b)***

**Correction:**

1- Losses are proportional to conductivity g and the square of thickness a, which calls for the use of low-conductivity, thin-walled sheets.

2- Since the conductivity of steel is lower than that of aluminum, steel sheets must be used; numerical application leads to :

$\boxed{p_J(\text{acier}) = 52{,}1\ W.m^{-3}}$ et $\boxed{p_J(\text{aluminium}) = 6100\ W.m^{-3}}$

# Part 5:
# Controlled by :
# Ultra Son

## Exercise 11: Ultrasonic non-destructive testing

1- A 200 mm thick steel shim is tested, with a significant discontinuity at 150 mm (Fig. 14).

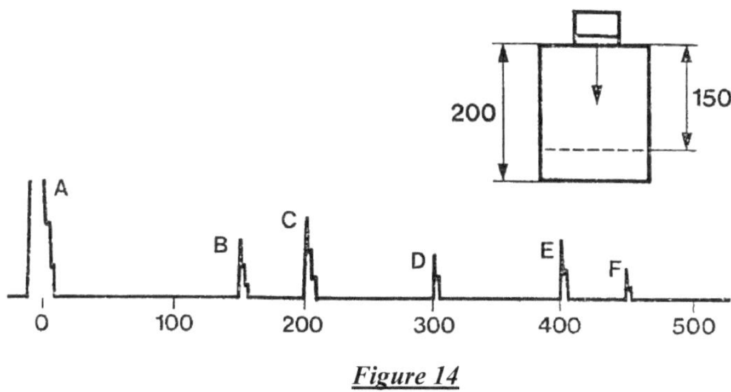


*Figure 14*

1-a What do indications A, B and C represent?

1-b Indication D corresponds to the second fault echo. What do indications E and F represent?

2- The figure below shows an immersion test on a 75 mm thick steel shim with a discontinuity located 50 mm from the entry face (Fig. 15) :

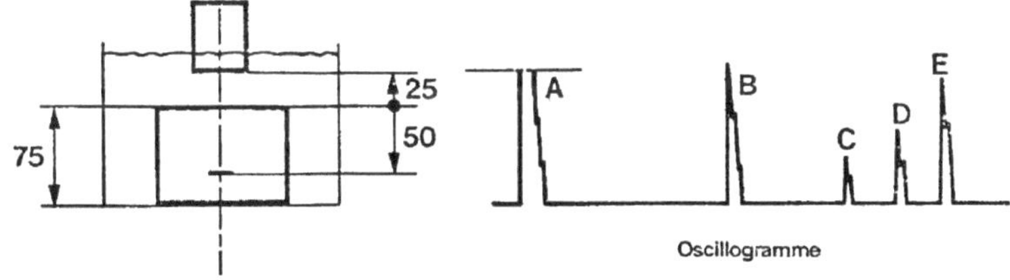


*Figure 15*

Indicate the position on the oscillogram:

- of the emission echo
- the echo of the defect

- from the echo in the background

Ultrasonic velocity in water 1500 m/s - Ultrasonic velocity in steel 6000 m/s

**Correction:**

1- The indications A, B and C in the image probably correspond to echoes detected during an ultrasonic inspection of a steel wedge. Here's one possible interpretation:

**Indication A :**

- Part **bottom echo** (or surface echo) detected at the start of the scan at 0 mm.
- This is the echo of the signal reflected by the **upper surface of** the part (200 mm in total thickness).

**Indication B :**

- This echo appears at around **150 mm** on the horizontal scale.
- It could represent **the internal discontinuity** located 150 mm from the surface. It may represent a significant defect in the part (crack, inclusion, cavity, etc.).

**Indication C :**

- This echo is observed at around **200 mm**, which corresponds to the total thickness of the part.
- **This is the echo from the bottom of** the hold, i.e. the signal reflected from the **underside**.

In a nutshell:

- **A**: Surface echo.
- **B**: Echo of the internal discontinuity at 150 mm.
- **C**: Bottom echo (200 mm).

These indications help locate the defect in relation to the total thickness of the shim.

2- **System and oscillogram analysis**

- **Part thickness**: 75 mm
- **Discontinuity** located 50 mm from the **entrance face** (the side from which the ultrasound signal enters).
- **Ultrasonic speed** :
  - **In water**: 1500 m/s

- **In steel**: 6000 m/s

The oscillogram displays several echoes corresponding to different interfaces encountered by the ultrasonic wave.

**Elements to identify on the oscillogram**

1. **Emission echo (A)** :
    - This is the **first peak**, representing the start of the signal emitted by the transducer. It corresponds to the moment when the wave enters the water and propagates towards the room.
2. **Fault echo (C)** :
    - This peak appears at the position corresponding to **50 mm** depth in the wedge.
    - The travel time of the ultrasonic wave between the input and the defect depends on the propagation speed in the steel:

$$t = \frac{2 \times 50}{6000} = \frac{100}{6000} = 0,0167\,\mathrm{ms}$$

    - This indicates that this echo represents the **internal discontinuity** detected at 50 mm.
3. **Echo from the bottom (E)** :
    - The last peak corresponds to the wave that crossed **the entire thickness of** the hold and was reflected by the **bottom** (75 mm).
    - Total travel time is :

$$t = \frac{2 \times 75}{6000} = \frac{150}{6000} = 0,025\,\mathrm{ms}$$

    - This allows us to identify this echo as the **return from the bottom**.

**Position of echoes on the oscillogram**

- **A**: Emission echo.
- **C**: Fault echo at 50 mm.
- **E**: Bottom echo at 75 mm.

These echoes can be used to locate and characterize the internal discontinuity at 50 mm.

## Exercise 12: Non-destructive testing: transverse wave corner probe

For a sufficiently small angle of incidence, a double refraction phenomenon is observed: a longitudinal wave a transverse wave are formed in the steel.

To obtain only the transverse wave in steel, the longitudinal wave can be suppressed by total reflection (Fig. 16).

1- What is the condition for total reflection of longitudinal waves at the angle of incidence i?

2- Calculate i so that the refraction angle of the transverse waves is 70°.

We give: c steel L = 5900 m/s, c steel T = 3200 m/s and c plexiglass L = 2700 m/s.

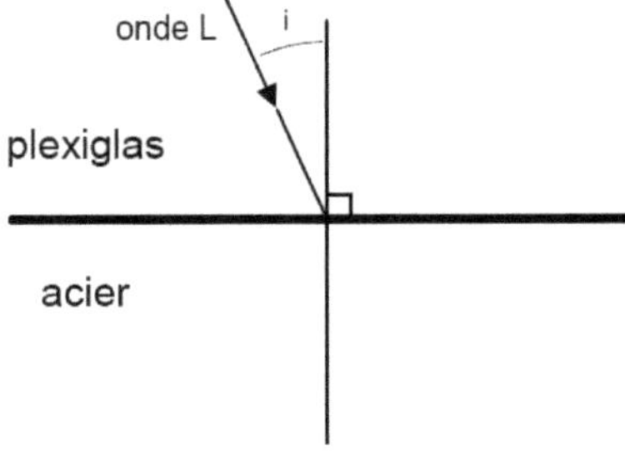


***Figure 16***

**Correction:**

1. Total reflection condition for longitudinal waves

Total reflection of a longitudinal wave occurs when the corresponding refraction angle reaches 90°. At this point, the entire longitudinal wave is reflected, and there is no longer any transmission in the material.

Snell's law is given by :

$$n_1 \sin(i) = n_2 \sin(r)$$

i is the angle of incidence,

r is the angle of refraction,

$n_1$ and $n_2$ are the refractive indices, related to the wave velocities in the two media by

$$n = \frac{c_{\text{milieu de référence}}}{c_{\text{milieu considéré}}}$$

For total reflection, we want the angle of refraction for the longitudinal wave to be 90° . Thus :

$$\sin(90^\circ) = 1 \quad \Rightarrow \quad n_1 \sin(i_{\text{limite}}) = n_2$$

In our case :

Input medium: plexiglass ($C_{\text{plexiglass L}}$=2700 m/s),

Output medium: steel ($C_{\text{steel L}}$=5900 m/s).

Thus :

$$\frac{c_{\text{plexiglas L}}}{c_{\text{acier L}}} = \frac{2700}{5900} = 0,4576$$

Therefore, the limiting angle of incidence for total reflection is :

$$\sin(i_{\text{limite}}) = 0,4576 \quad \Rightarrow \quad i_{\text{limite}} = \arcsin(0,4576) \approx 27,2^\circ$$

2. Calculate the angle of incidence so that the angle of refraction of transverse waves is 70°.

Here, we want the transverse wave in the steel to have a refraction angle $r_T$=70°.

Again, we use Snell's law:

$$n_1 \sin(i) = n_T \sin(r_T)$$

$$n_1 = \frac{c_{\text{plexiglas L}}}{c_{\text{acier T}}} = \frac{2700}{3200} = 0,84375,$$

$$r_T = 70^\circ.$$

We then calculate :

$$\sin(i) = 0,84375 \times \sin(70^\circ)$$

$$\sin(i) = 0,84375 \times 0,9397 \approx 0,7924$$

So :

$$i = \arcsin(0,7924) \approx 52,3^\circ$$

## Exercise 13: Variable angle

Consider a variable-angle transducer consisting of a piezoelectric blade fixed along a diametral plane of a Plexiglas cylinder (see figure showing a cross-section of this transducer).

This cylinder rotates, fitting snugly into a hole in the parallelepipedic part that makes up the second part of the transducer, also made of Plexiglas. A coupler provides what we consider perfect transmission. If we call a the angle between the plane containing the piezoelectric blade and the horizontal plane (see figure 17).

1- Determine the range of variation of a that allows only transverse waves to be generated in a steel part.

***Question***: *What angle do I have to hang to get 45° transverse waves in the same piece of steel?*

The following numerical values are given:

Speeds in steel: $V_L = 5850$ m/s; $V_T = 3230$ m/s.

Speeds in Plexiglas: $V_L = 2730$ m/s; $V_T = 1430$ m/s.

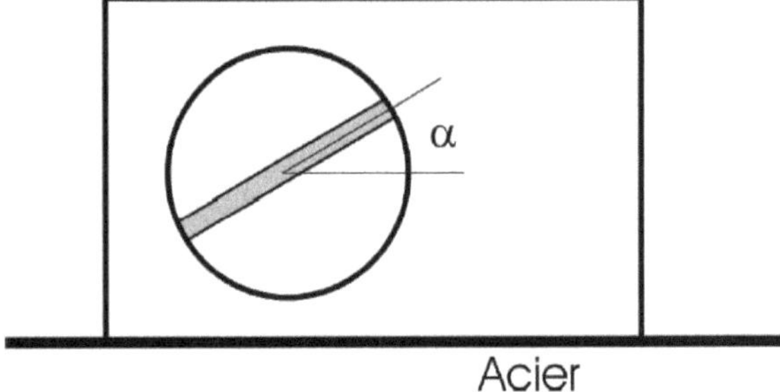


***Figure 17***

**Correction:**

1. Determine the range of variation of the angle α to generate only transverse waves in the steel.

Principle: Total reflection of longitudinal waves

To generate only transverse waves in steel, the longitudinal wave must be totally reflected at the plexiglass-steel interface. This means avoiding the transmission of longitudinal waves in steel, which occurs when the refraction angle for longitudinal waves reaches 90°.

Total reflection condition for the longitudinal wave :

$$\sin(i_{\text{limite}}) = \frac{V_{\text{plexiglas L}}}{V_{\text{acier L}}}$$

With :

$V_{\text{plexiglas L}} = 2730\,\text{m/s},$

$V_{\text{acier L}} = 5850\,\text{m/s}.$

$$\sin(i_{\text{limite}}) = \frac{2730}{5850} \approx 0,4667$$

$$i_{\text{limite}} = \arcsin(0,4667) \approx 27,8^\circ$$

Therefore, the angle α must be greater than 27.8° to avoid longitudinal waves being transmitted into the steel. This ensures that only transverse waves are obtained.

2- We now want the refraction angle for transverse waves in steel to be 45°.

Application of Snell's law to transverse waves :

$$\sin(i) = \frac{V_{\text{plexiglas L}}}{V_{\text{acier T}}} \times \sin(45^\circ)$$

With

$V_{\text{plexiglas L}} = 2730\,\text{m/s},$

$V_{\text{acier T}} = 3230\,\text{m/s}.$

$$\sin(i) = \frac{2730}{3230} \times 0,7071 \approx 0,5978$$

$$i = \arcsin(0,5978) \approx 36,8^\circ$$

So, to obtain transverse waves with an angle of 45° in steel, the angle of incidence α must be 36.8°.

## Exercise 14: Measuring the thickness of a material

A 10 mm thick steel plate is tested with an ultrasonic transducer at a given frequency.

Data : Speed of sound in steel = 5900 m/s.

1- Calculate the total time of flight (round trip) of the ultrasonic wave.

2- If a 2 mm reduction is observed, determine the new thickness.

**Correction:**

1- Calculate the total time of flight (round trip) of the ultrasonic wave:

Data: Thickness = 10 mm = 0.01 m, Speed of sound = 5900 m/s.

Total time of flight = (2 x Thickness) / Speed = (2 x 0.01) / 5900 = 3.39 μs.

2- Determine the new thickness after reduction:

New thickness = Initial thickness - Reduction = 10 mm - 2 mm = 8 mm.

## Exercise 15: Detecting internal faults

A test block contains an internal crack at an unknown depth.

Data: Measured time of flight = 50 μs; propagation speed = 3200 m/s (aluminum alloy).

Calculate the distance of the crack from the surface.

If the wave returns within 120 μs, determine whether another anomaly is present.

**Correction:**

1- Calculate the distance of the crack from the surface :

Data: Measured flight time = 50 μs = 50 x $10^{-6}$ s, Speed = 3200 m/s. Distance = (Time of flight x Speed) / 2 = (50 x $10^{-6}$ x 3200) / 2 = 0.08 m = 8 cm.

2- Check for any other anomalies:

Total time = 120 μs, so new distance = (120 x $10^{-6}$ x 3200) / 2 = 19.2 cm. Another anomaly is detected at this depth.

## Exercise 16: Influence of frequency on detection

Objective: To study the impact of ultrasonic frequency on resolution and penetration.

Material: aluminum; thickness = 20 mm.

Hypothesis: Test at 2 MHz, 5 MHz, and 10 MHz.

1- Compare the results obtained in terms of resolution and depth achieved.

2- Explain how to choose the optimum frequency for this material.

### Correction:

1- Comparison of results :

- At 2 MHz : Good penetration, low resolution.
- At 5 MHz: Compromise between resolution and penetration.
- At 10 MHz: Excellent resolution, low penetration.

2- Optimum frequency :

For a thickness of 20 mm, choose 5 MHz for a compromise between penetration and resolution.

## BIBLIOGRAPHY


- Normalisation française A 09-325 Essais non destructifs Ultrason Faisceau acoustiques – Généralités Septembre 1987.

- Soudage et techniques connexes, Recueil de Normes Françaises AFNOR A 09-320, contrôles et essais, Tome 3, 1984.

- Des contrôles ultrasonores de joints soudés austénitiques H.W. Corsepius Krautkramer France.

- Le contrôle non destructif par ultrasons, Jean Perdijon, Traité des nouvelles technologies Hermes Science Publications, 1993.

- L'émission acoustique, cours de contrôle et d'inspection des constructions soudées, Institut de Soudure, ESSA 1984.

- Manuel pour l'examen par ultrasons des soudures, Institut international de la soudure, Publications de la soudure Autogène, 1978.

- Contrôle des constructions soudées, contrôle par ultrasons, Institut de soudure, ESSA, 1983.

- Le contrôle non destructif et la contrôlabilité des matériaux et structures, Gilles Corneloup, Cécile Gueudré ; Presses polytechniques et universitaires Romandes, 2016.

- Guide ultrasons multiéléments, Principes et applications pour le contrôle non destructif, François Berthelot – Benoit Dupont, 2013.

- Contrôle non destructif (CND) Jacques Dumont-Fillon, Techniques de l'Ingénieur, Traité Mesure et Contrôle.

- Technologie des métaux, contrôles et essais des soudures, Michel Bramat (Mayer, Villeneuve), de boeck, 2008.

- Ultrasons de haute intensité, applications industrielles, B. Brown et J.E. Goodman, Dunod, 1971.

Printed by Books on Demand GmbH, Norderstedt / Germany